*For Hugo, Tala,
and Ben*

Spark an Interest: The Green Alphabet Discovering Sustainability from A to Z*

For permission requests, write to the author at rachelreyesbrandwood@gmail.com.

NATIONAL LIBRARY OF AUSTRALIA

A catalogue record for this book is available from the National Library of Australia

ISBN: 978-1-7640267-8-9 (paperback)
ISBN: 978-1-7640267-9-6 (eBook)

Spark*an Interest

The Green Alphabet: Discovering Sustainability from A to Z

Rachel Reyes-Brandwood

Sustainability

means taking care of the **planet**, **people**, and **resources** so that everyone can live well now and in the future. It's about protecting nature (**environmental**), helping communities thrive (**social**), and making smart choices with wealth (**economic**) so nothing runs out.

A is for Alternative Energy

ID 97448690 © Airubon | Dreamstime.com

Wind turbines and solar cells generate alternative renewable energy.

Alternative energy comes from sources like the sun, wind, and water instead of fossil fuels. These energy sources are cleaner and better for the planet because they don't pollute the air. By using alternative energy, we can power our world while protecting nature.

B is for Biodiversity

ID 7626493 © Debra James | Dreamstime.com

Coral reefs and fishes at the Great Barrier Reef are threatened by biodiversity loss.

Biodiversity means having lots of different plants, animals, and insects living together. Each species, from bees to coral reefs, plays an important role in keeping ecosystems healthy and balanced. Protecting biodiversity helps ensure a thriving planet for everyone.

C is for Carbon Footprint

ID 244470976 © Anna Bergbauer | Dreamstime.com

Burning fossil fuels for electricity, heat, and transport is the largest source of human-caused carbon emissions.

Carbon footprint measures how much greenhouse gas you create by doing things like driving or using electricity. The smaller your footprint, the better it is for the planet! Turning off lights when you leave a room, or walking instead of driving short distances, are easy ways to shrink your footprint.

D is for Deforestation

ID 354005186 © Branislav Kapetanovic | Dreamstime.com

Logging contributes to climate change.

Deforestation happens when too many trees are cut down, often to make space for farms or buildings, or harvest wood. Trees are important because they clean the air and provide homes for animals. We can help forests by planting new trees and recycling things like paper.

E is for Ecosystem

Shore birds avocets and dowitchers capture aquatic insects in thriving wetland ecosystem, west of the Great Sand Dunes, Colorado, USA.

An ecosystem is a community of plants, animals, and their environment working together. Every part of an ecosystem depends on the others to survive. Keeping ecosystems healthy makes sure that plants, animals, and people have access to clean water, fresh air, and safe habitats for future generations.

F is for Fair Trade

Cocoa pods and cocoa beans are usually harvested by farmers who face tough challenges like low pay, hard work, and worries about bad weather or losing their crops.

Fair trade guarantees that farmers and workers get paid fairly for their hard work, that they work in decent conditions, and there's no forced labour, child labour, or discrimination. It also means products are made in ways that are kind to the planet. Next time you eat chocolate, look for the fair trade label. It means your treat gave farmers fair wages, and that they worked safely and farmed sustainably.

G is for Greenhouse Gases (GHGs)

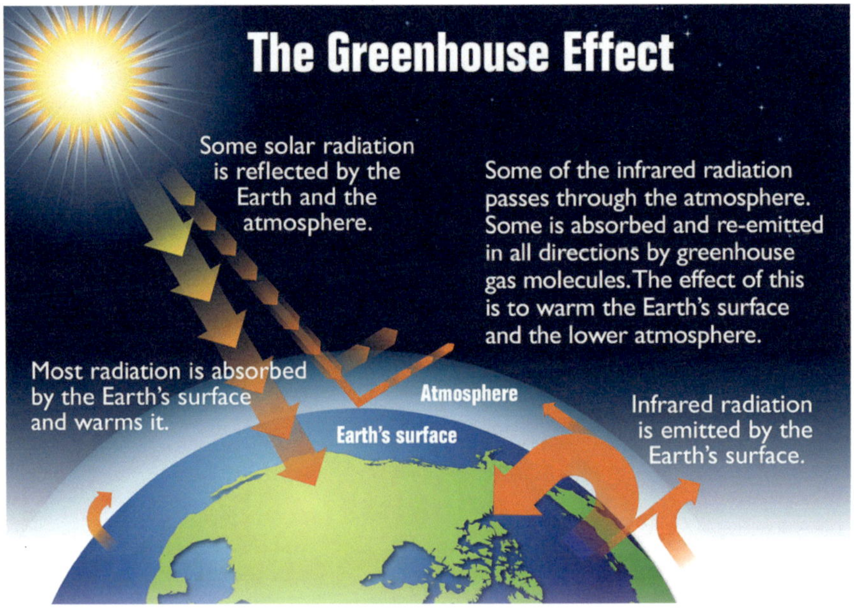

Greenhouse gases in the atmosphere trap heat from the sun and keep Earth warm enough for life to exist. But too much greenhouse gases, like carbon dioxide from cars and factories, methane from livestock, and nitrous oxide from fertilisers, cause global warming and climate change, leading to extreme weather events. Reducing these gases helps protect our planet's future by slowing down climate change and keeping Earth safe for everyone.

H is for Habitat Restoration

ID 72727642 © Seadam | Dreamstime.com

Mangrove roots serve as home to fish and jellyfish.

Habitat restoration means fixing damaged areas so plants and animals can live there again. It's like giving nature a second chance to thrive! For example, planting mangroves along coasts protects shorelines and creates homes for fishes and birds.

I is for Input-Output Analysis

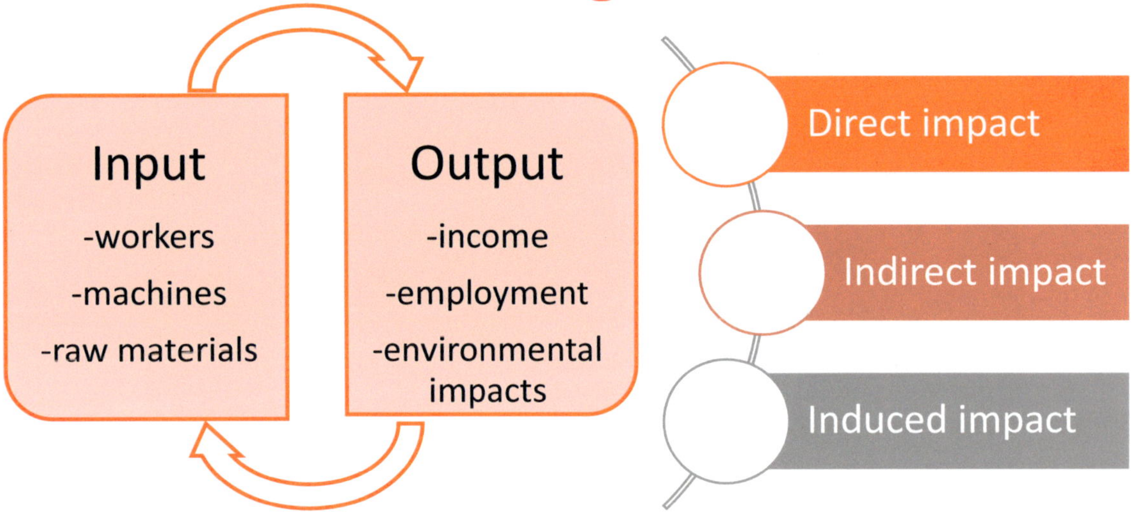

Input-output analysis shows how different parts of the economy—like farms, factories, and stores—are all connected and dependent on each other. By looking at these connections, we can see how making or changing something (like building a bridge or growing food) affects money, people, and the environment, even in surprising ways. This lets us understand the full impact of our actions so we can do things that are better for the planet, people, and the economy.

J is for Just Transition

ID 344020014 © Korn Vitthayanukarun | Dreamstime.com

Retraining workers from traditional polluting industries like coal can help them secure green jobs, such as working on wind turbines.

A just transition means making changes to protect the planet in a way where nobody is left out. It's about helping people find new, greener jobs and making sure everyone's voice is heard. Countries with more money help those with less, so the transition is fair for everyone.

K is for Kyoto Protocol

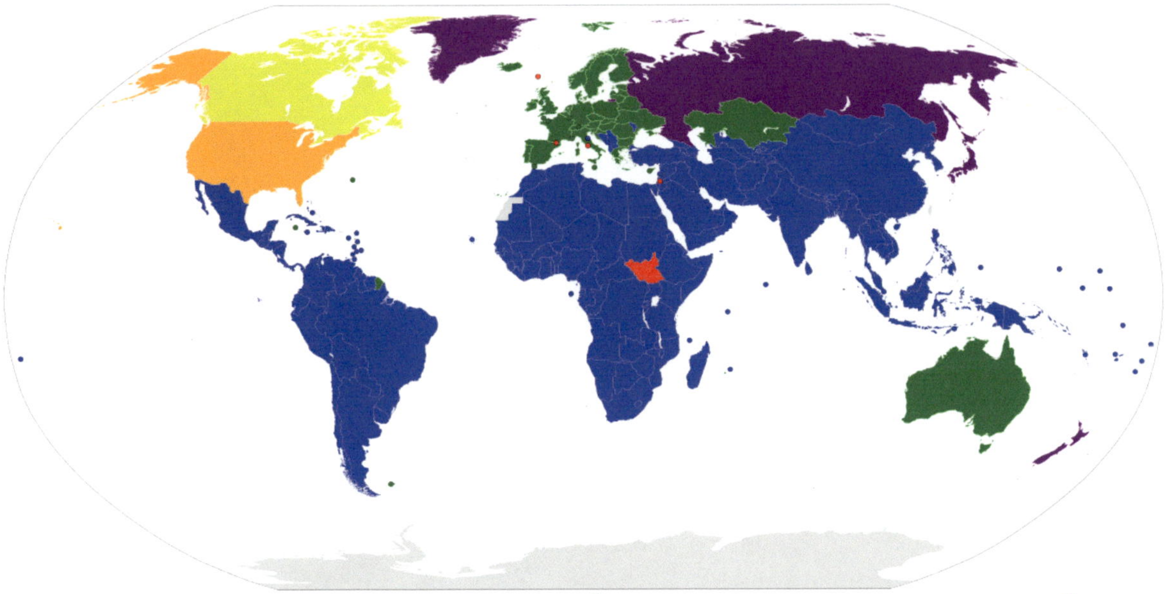

Map of participation to the Kyoto Protocol: Countries with binding targets, 2012-2020 Countries with binding targets, 2008-2012 Countries with no binding targets Countries which withdrew from the Protocol Countries that did not ratify the Protocol Other UN member states and observers

The Kyoto Protocol was an agreement between countries to reduce greenhouse gas emissions. It was one of the first big steps toward fighting climate change together. Even though not all countries had to follow the same rules and some didn't join at all, it helped start the idea of teamwork for sustainability and inspired newer agreements. (See letter P.)

L is for Low-Carbon Living

ID 234983834 © Milos Ruzicka | Dreamstime.com

A houseboat with solar panels on the Amstel River in the Netherlands is an example of low-carbon living.

Low carbon living means making choices that create less carbon emissions, like taking public transport instead of driving, and eating more veggies and less meat. Every little action adds up to help the Earth! It's about living simply while caring for nature.

M is for Modern Slavery

ID 315907306 © Mamunur Rashid | Dreamstime.com

The Rana Plaza collapse in Bangladesh in 2013 exposed dangerous and unfair working conditions in the global fashion industry.

Modern slavery happens when people are forced to work without fair pay or freedom. Sometimes, they are made to work to pay off debts, or are tricked or taken to other places. Choosing ethical products helps fight this injustice and promotes sustainability.

N is for Net Zero Emissions

ID 258843920 © Paradee Paradee | Dreamstime.com

Net zero long-term strategies include energy efficiency, use of clean energy, carbon footprint reduction, and adoption of carbon capture technology.

Net zero emissions mean balancing the greenhouse gases we create with actions that remove them, like planting trees or using clean energy. Countries like Australia have set net zero as a goal by 2050, working together to reduce pollution and protect the planet. Reaching net zero helps slow down climate change and paves the way to a healthier future for everyone.

O is for Organic Farming

ID 35261040 © Denise P. Lett | Dreamstime.com

Bunches of colourful organic rainbow carrots are displayed for sale in a market.

Organic farming involves growing food without harmful chemicals that hurt soil, water, or wildlife. It's healthier for people and kinder to the planet. Choosing organic food supports sustainable agriculture.

P is for Paris Agreement

The Paris Agreement is a promise made by nearly every country around the world to work together to fight climate change. It aims to keep Earth's temperature rise below 2°C, and ideally under 1.5°C, to protect nature and people. Every country sets its own goals to reduce pollution and helps each other with money and technology to make the planet healthier.

Q is for Quality of Life (Wellbeing)

Quality of life means feeling happy, healthy, and safe in your community. We promote wellbeing by protecting nature, creating fair opportunities for everyone, and ensuring access to education so people can learn and grow. A safe planet with happy and healthy people means better lives for all!

R is for Reduce, Reuse, Recycle (3 R's)

ID 244228082 © Anna Bergbauer | Dreamstime.com

The 3 R's help us use less stuff and create less waste! Reduce what you use, reuse what you can, and recycle materials into new things. Following these steps keeps our planet clean and green.

S is for Sustainable Development Goals (SDGs)

The SDGs are 17 global goals to end poverty, protect the planet, and ensure fairness by 2030. They guide countries in making smarter choices for people and nature, but we still have a lot to do. Next, countries will check their progress and make new plans to keep making things better for everyone.

T is for Triple Bottom Line

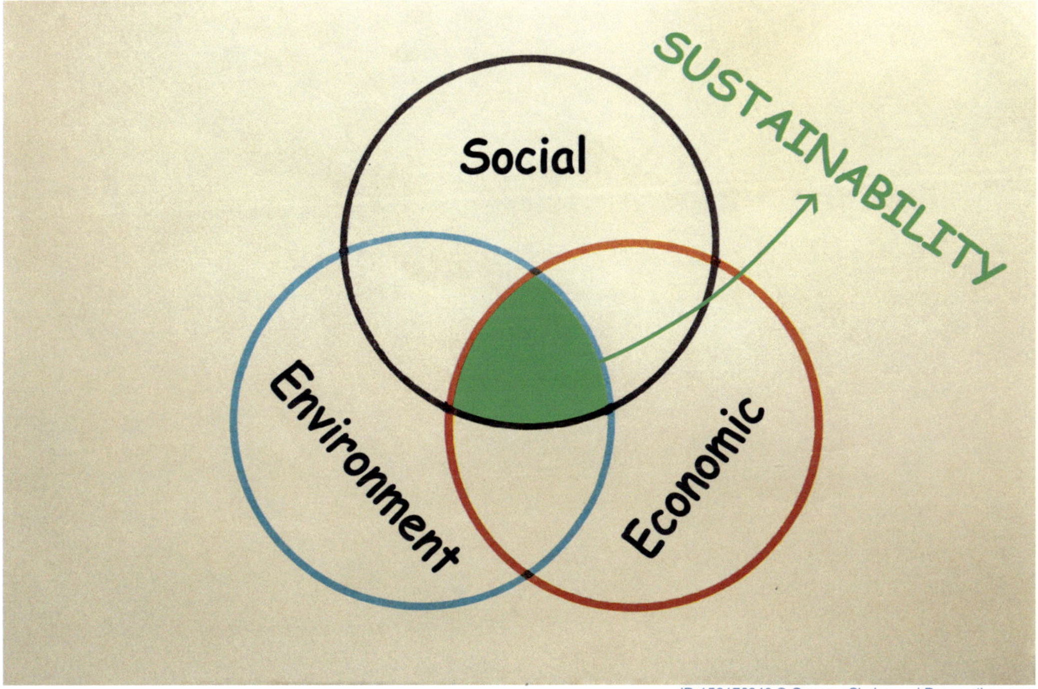

Balancing social, environmental and economic factors leads to true sustainability.

The triple bottom line balances three things: people (social), planet (environmental), and profit (economic). Businesses use this idea to grow responsibly without harming communities or nature. It's about success that benefits everyone!

U is for Urban Planning

Singapore has pioneered the integration of green spaces, sustainable infrastructure, and nature-based solutions into its dense urban landscape.

Urban planning means designing cities with green spaces, clean transport, and smart buildings. Good planning creates places where people can live happily while protecting nature. Think bike lanes, rooftop gardens, and solar-powered charging stations!

V is for Value Chain Sustainability

Sustainable material sourcing, production, delivery, consumption and recycling

Value chain sustainability means making sure every step in creating a product—from materials to delivery—is eco-friendly and fair to workers. It reduces waste and pollution while supporting communities along the way. Sustainable choices make a big difference!

W is for Water Footprint

ID 35350650 © Symbiot | Dreamstime.com

Beef is the food product with the highest water footprint because of the water needed to grow feed for cattle. Producing a kilogram of beef requires over 15,000 litres of water!

Water footprint measures how much water you use daily through activities like drinking or growing food. Saving water helps protect this precious resource! Small changes like turning off the tap while brushing your teeth can make a big impact.

X is for Xeriscaping

Los Angeles Air Force Base uses xeriscaping to conserve water.

Xeriscaping uses plants, including native ones, that need little water to create beautiful gardens. It's perfect for dry areas, saves water while supporting local wildlife, and an example of working with nature. Cacti, lavender, and succulents are xeriscaping superstars!

Y is for Yield

ID 184802405 © Rodrigolab | Dreamstime.com

The Ifugao Rice Terraces at Batad, Philippines is a model of sustainable agriculture.

Yield means how much food or energy we produce from resources like land or sunlight. Sustainable practices make sure we get enough without harming nature or wasting resources. A good yield benefits both people and the planet!

Z is for Zero Waste Lifestyle

ID 185592775 © Hollyharryoz | Dreamstime.com

Compost from kitchen waste improves soil fertility in gardens and vegetable beds.

Zero waste living means creating no trash by reusing and recycling everything possible. It's about being creative with what you have instead of throwing things away! You can compost food scraps, use reusable straws, and swap toys with friends, to help keep our Earth clean and with resources for future generations.

Made in United States
Troutdale, OR
11/03/2026

56844674R00021